The Grand Beehive

Compiled and with an introduction by
Hal Cannon

Designed by Brian Jones
Photographed by Brent Herridge
Produced by Skip Branch

University of Utah Press
Salt Lake City 1980

Foreword

The beehive is, among many other things, the symbol of the State of Utah. It is perhaps the most widely used of all state symbols, whether animal, vegetable, or mineral. In Utah there are representations of beehives everywhere, a fact of which I was only subconsciously aware before Hal Cannon and David Pendell visited me to propose an exhibition devoted to exploring the use of the beehive image. What first seemed a delightful but ridiculous idea began to appear more and more worth pursuing. There *are* beehives everywhere in Utah. The realization has led to what I hope is a short-term obsession and a long-term awareness for the artists, craftsmen, and all who have helped to produce the beehive exhibition.

This book is an extension and to a degree a record of the Grand Beehive Exhibition, presented at the Salt Lake Art Center from September 12 through October 19, 1980, and in another version at the Renwick Gallery of the Smithsonian Institution in Washington, D.C., from April 24 until December 6, 1981. The exhibition contained many more objects than does the book. Some things were created especially for the show, other items were "found," and they range from the beautiful to the bizarre. The correspondence art component of the Grand Beehive Exhibition, which has only a small representation here, included over a hundred diverse and often witty postcard interpretations of the beehive theme from artists all over the world. Besides the works made for the show, we found beehives in quilts, neon signs, gravestones, buildings, lapel pins and shoulder patches, fire hydrants, newel-posts, pop bottles and sidewalks.

The myriad, wonderful Beehives of Utah may or may not outnumber other states' Keystones, Poppies, Potatoes, and Lone Stars, but they certainly are a joy to see. Beehives are everywhere.

ALLEN DODWORTH
DIRECTOR, SALT LAKE ART CENTER

at the Renwick Gallery of the
Smithsonian Institution in Washington, D.C.
from April 24 until December 6, 1981. The
exhibition contained many more objects than
does the book. Some things were created
especially for the show, other items were
found, and they range from the beautiful to
the bizarre. The more spectacular art con-
tent of the Craft became Exhibition which
has a small representation here, included over a hundred objects and often were
idiosyncratic interpretations of the beehive theme
from artists all over the world. Besides the

Acknowledgments

The Grand Beehive Exhibition and Catalog
came about with the generous asssistance of
  Utah Arts Council
  Mervyn's Stores
  Salt Lake Art Center
  Renwick Gallery of the National Collection
    of Fine Arts, Smithsonian Institution
  National Endowment for the Arts

Beehives are everywhere in Utah. But familiarity breeds transparency, and today the beehives rarely are noticed. Since this project began many people have noticed beehives, have searched them out, and have advised us about them. Others have let us borrow and photograph their beehives. And still others have helped us catalog, display, and publish our findings. To all these people who have joined us in the cult of the beehive, our gratitude is extended.

Special thanks go to those who assisted in organizing this effort, among them: Ruth Draper, Carol Edison, and the board and staff of the Utah Arts Council; Allen Dodworth, D. A. Hardy, and the staff and trustees of the Salt Lake Art Center; David Pendell and Utah Designer Craftsmen; Paula Roberts and Norma Mikkelsen, the University of Utah Press; Bess Lomax Hawes and the staff of the Folk Arts Program at the National Endowment for the Arts; Congressman Gunn McKay and his staff; Governor Scott M. Matheson, Lieutenant Governor David S. Monson, and the Department of Community and Economic Development; and Lloyd Herman, Ellen Myette, Elaine Eff, Val Lewton and the staff of the Renwick Gallery of the National Collection of Fine Arts.

Thanks also go to those who helped in researching the historical beehive: Austin E. Fife, Utah State University; Allen D. Roberts, *Sunstone*; Tom Carter, Jay Haymond, and Tim Neville, Utah State Historical Society; Dennis Rowley, Brigham Young University; Eva Crane, International Bee Research Association; Roger A. Morse, Cornell University; Frederick M. Huchel, Brigham City Museum Gallery; Nancy Richards, Pioneer Trail State Park; Richard Oman, Jim Kimball, and Paul Anderson, Church of Jesus Christ of Latter-day Saints; Jonathan L. Fairbanks, Museum of Fine Arts, Boston; Elly Schoenfeld, Salt Lake City *Deseret News*; Larry Jackstien, Hotel Utah; Everett Cooley, University of Utah; Ruth Garbett and Dorothy Reeve, Daughters of Utah Pioneers Museum; and John Sillito and Pat Sullivan, research assistants.

Additional thanks go to those who assisted with the preparation of the exhibition and catalog: Bruce Baker, David Blaine, Hal Cannon, Carol Edison, Hal Rumel, and Terry Newfarmer, contributing photographers; Hugh C. Bringhurst and Claud Montier, Utah State Exhibitions; John McClay, Utah State Archives; Florence Jacobsen, Bernice Casper, Glen Leonard, Linda Gibbs, Bill Slaughter, Stephen

Fletcher, Jay Rosenberg and Warren Luch,
Church of Jesus Christ of Latter-day Saints;
Myron Barlow, Seven-Up–Dr. Pepper Bottling
Co., Ogden; Sandy Metal Works; Utah|Auto
Wrecking; Erwin Farnsworth, C. Laird
Snelgrove Jr., and Elaine Wassmer, Snelgrove
Ice Cream Co.; Tom Young Jr., Young Electric
Sign Company; Sam Weller, Meg Brady, Alice
Edison, Lynette Hart, Gail Madden, Calvin and
Lucybeth Rampton, Glenn Richards, Wanda
Wade, and Allan Wardle.

A reading of the written material contained in
this catalog has been provided for the visually
impaired and is available at the Utah State
Library in Salt Lake City.

All proceeds from the sale of this catalog will
be used to fund additional Utah folklife projects.

The Beehive in Utah Folk Art

The people of Utah have inherited and preserved a remarkable and largely unexamined treasury of folklore and art. An attractive aspect of this inheritance is the unusual way in which one particular traditional motif appears repeatedly. The recurrent emblem is the beehive, and its use is both striking and pervasive. More importantly, the beehive motif carries with it a wealth of meaning and information about a people's cultural and psychological connections with the distant past and with its immediate and unique history.

The Beehive in Use and Symbolism

The beehive, called by English beekeepers a "skep," is a woven half-sphere with a long history of use in northern Europe which reaches back to the Middle Ages. The beehive's purpose, of course, is to house bees and make their honey accessible to men. If we were to find a rare one in use today it would be in northern Europe on a farm run in the traditional manner, where honey was not the cash crop. In many countries the skep has been outlawed. While utility is a primary characteristic of the beehive, it has always functioned as much more than a farm implement, for it has also been invested by man with symbolic meaning. We cannot really separate the beehive as mythic emblem from the beehive as utilitarian object, since the history of the relationship of bees to man is at the core of man's beehive mythology.

As primitive cultures around the world became acquainted with the bee they found a mysterious animal which was at once a docile provider of sweetness and a stinging enemy. This natural paradox required solution, and the cultures provided myths to bring harmony to the discordance. Throughout the history of the myth of the beehive, importance has been attached to it at both the personal and symbolic levels.

A. C. Lambert, in his study of symbology, tells us that one of the most inclusive prescriptions of meaning, the Rosicrucian system, like many Western mystic systems such as Masonry, belief interpreted the hive as a place for working and building. It represents a tabernacle inside which we work together in concert. In our work we are servants to others and we take from nature for the good of our fellow man. Honey represents the goal of that quest. We must also take elements from the world to build a sound temple for our spirit to reside within, just as the bee builds a hive in order to store its honey.

The History of the Hive

When early man first found the hive, honey bees were already being robbed by ancient birds and bears. By the time of the mighty Egyptian dynasties, bees were being cultivated in hollow logs, clay pots, and mud and straw apiaries. Already the bee, its life, its products (honey and wax), and its harmonious community life were observed and idolized.

The bee appeared prominently in Greek literature, most importantly in the work of Virgil, who described bee attributes in poetry and so embodied a history of human awe of this small beast. Significantly, bees dwelt harmoniously in colonies of up to two hundred thousand individuals, gave wax which was the earliest of sculpted materials and whose light shone on holy rites, and supplied honey which was not only the most pleasant of food and a potent medicine, but which, when fermented and drunk, freed the soul with some unknown magic. It is easy to imagine how the speculative mind might have interpreted the behavior of this little animal whose mysterious life was protected by a sting and cloistered from view in a hollow log. Even Aristotle, in correctly describing the procreation of most animals, speculated on the mystery of the ways in which bees might have been created and in turn recreated their kind. (He put forward a fairly accurate description which took into account the queen bee, but he ultimately dismissed the hypothesis as too fantastic to be tenable.) He finally concluded that the bee, unlike wasps and hornets, must be of divine origin.

Belief in the divinity of the bee is widespread in ancient culture; the number and variety of civilizations which have revered the bee is staggering. Austin Fife contends that no other animal life has had more mythological attention than the bee; many stories survive as evidence of its importance. Zeus hovered over Olympus and counseled Hermes to consult the bees for the future. Mohammed in the Koran relates that the Lord spoke directly to the bee, something he never did with any other animal. In some Hindu drawings, the bee represents heaven; in a Hindu fertility rite, a maiden's generative organ is anointed with honey. Asvins, lord of brightness, is beseeched in historic India to anoint men with honey so their speech will be more persuasive. According to the Mishra, beehives adorned the vestments of the priests of ancient Judaism.

Early Christianity inherited a tradition of bee worship which it incorporated into its own sacred mythology. We find early Christian accounts of the bee as a symbol of the chaste and the wise, as the foreteller of weather, the future, and the fate of armies, and also as the embodiment of the soul of man. Speculation on the bee's origin proposed that it was generated from heaven, from the corpse of an ox or lion, from the sun, or from the tears of Christ.

Almost every western culture has looked on the hive as a model of the perfect social system. Political theorists and monarchs throughout history have compared their systems of government to that of the beehive. Austin Fife has described the hive's political system: the bee colony has an absolute ruler who is obeyed by all. There is perfect division of labor and a well-organized military system which includes guards at the hive entrance. The bees are remarkably industrious and have a well-planned food storage system. The bee is a genius at architecture. The colony is capable of expansion; new colonies are organized regularly. With these admirable attributes, it is little wonder that the beehive became an emblem of perfect social order to the medieval Catholic church. European monarchs, as well as Napolean's armies, employed the beehive as a prominent symbol for political organization and power.

In European folklife bees and their hive were treated with reverence. Beekeepers used charms to keep bees from leaving the hives. Wooden hives were painted with religious scenes in historic Austria. Hives were decorated with ribbons at celebrations. Deaths in the family were announced to the bees as a first order of business. Folklore maintained that bees sang to honor Christ on Christmas Eve, and that bees renounced the pleasures of love out of respect to God.

Beehives in the Americas

Ancient Americans in the tropics of South and Central America cultivated wild honey from the stingless bee, which still flies undomesticated in those climes. When Spanish explorers visited Mayan temples they found a flourishing bee industry and accompanying lore. In North America, however, there are no records of an indigenous honey bee. The old world honey bee was shipped to Cuba in the seventeenth century where before long beehives appeared in every community of new world settlement. Both beekeeping and bee mythology paralleled the old world inheritance during the colonial period.

The hive as a symbol declined in the nineteenth century. The gradual replacement of honey with sugar and the replacement of honey-based alcohol with other intoxicants helped to lessen the bee's practical importance. Beekeeping drastically changed in the latter half of the nineteenth century as the old-style skep hive was gradually replaced by the new movable frame boxes, breaking the link between utility and myth.

The scientific approach to beekeeping has ended the use of the skep in America. Science has also meticulously observed the subtleties of the bees' life, so that the bee no longer seems a mysterious and pure animal sent to earth from heaven above. Although we still admire the bee's industry and enjoy its honey and wax, the bee has lost its status as a sacred animal.

Mormons, Masons, and the Beehive

One hundred and fifty years ago Joseph Smith, a young man from upstate New York, established a religion called the Church of Jesus Christ of Latter-day Saints. The church began with a small group of neighbors and family who were persuaded that Joseph Smith had received a vision which instructed him to re-establish the true church of Christ. The young prophet was told by an angel to dig up plates of gold on which appeared hieroglyphs. Through divine guidance he translated these plates into a volume called the Book of Mormon, which unfolded the history of the Americas. This new scripture along with the Bible formed the basis for the new religion, established at a time in American history of great religious revival. Like the writers of the Constitution, Joseph Smith and other key men in the new religion were Masons. During the early formulative period the Church found needed organization, ritual, and symbol in Grand Masonry. This adaptation was justified by the Mormons' conviction that much of the Masonic Order had survived in correct form from the true gospel of Christ.

Though the fledgling church conformed to a degree with the norms of nineteenth-century American life, its members were unusually zealous and frequently at odds with their non-Mormon neighbors. From Palmyra, New York, an exodus began which took the Mormons westward to three major settlements within the first fifteen years of the church's history. Each move was precipitated by conflict between the existing social norms and the Mormons' vision of life. Each move took them closer to the western frontier.

In Kirtland, Ohio, the Church outwardly resembled Protestant churches of the day. But in the second settlement at Nauvoo, Illinois, the holy symbols of Masonry, among them the sun stones, the square and compass, and the "all-seeing eye," were adopted and openly displayed on church buildings and publications and were employed as decorative motifs on a variety of objects. The use of these powerful and suspicious symbols was an additional provocation to the already hostile anti-Mormon residents of those areas.

When the Mormons built a city, Nauvoo, on the banks of the Mississippi River, Joseph Smith and his followers felt they finally had created a permanent home. But trouble with frontier neighbors led to the murder of Joseph Smith and his brother by a mob in 1844 as they were being held in the state's custody. The shock of their prophet's death was a final blow to the Saints of Mormondom who now found that their lovely city at the edge of civilization could not protect them. They would have to venture out into the wilderness to find a new home where they could live according to their tenets.

The Journal History of the Church of Jesus Christ of Latter-day Saints describes the burial of Joseph and Hyrum Smith after their martyrdom at the Carthage Jail. "The bodies remained in the cellar of the Nauvoo House…until the fall, when they were removed…at Emma's [Joseph Smith's wife] request to near the mansion…and the bee house was then moved and placed over the graves…." One might speculate that not only would hives of bees discourage grave molesters, but perhaps this act also expressed a belief that the bee is a benevolent link between man and the immortal soul.

The Bees of Deseret

The Mormon link to beehive iconography (of ancient tradition) came undoubtedly through Grand Masonry, but it is clear that in the building of a new Zion in Utah the beehive symbol came to have particular significance for the Mormons. Though we don't know exactly when or where the beehive took hold of the Mormon imagination, we can speculate that its importance grew soon after Brigham Young succeeded Joseph Smith and led the Church to the West. As the builder of the Kingdom of God on Earth, as general, and as candidate for President of the United States, Joseph Smith envisioned a destiny wherein America would embrace the concept of the Kingdom of God through a rapid conversion of the masses to the new religion. As the church recovered from the disappointment of this hope, the new prophet, Brigham Young, designed a different Kingdom, isolated, orderly, and independent, whose scope was more gradual. This was to be a place in which the forces of good could be generated for the eventual realization of a latter-day world-wide Kingdom. This vision, much more than Smith's, was very like the traditional kingdom of the hive. The beehive, of course, was built upon the symbols of the earlier church, but it surely represented Brigham Young's kingdom. The beehive began its course to becoming the most pervasive symbol of Mormondom at some time during the immigration of the Saints to a new Zion out on the great western wilderness.

That the importance of the beehive grew during the move west is supported by a suggestive parallel in Judeo-Christian tradition. The Mormons during this period of upheaval had always in mind the account in the Old Testament of the flight of Moses and the Israelites from their persecutors. Mormon church leaders comforted their followers with this comparison and promised a similar "land of milk and honey" at the eventual settlement in Zion.

A similar event in the Book of Mormon is the flight in ancient times of a group of righteous people called the Jaredites from the unholy city of Babel. God led the Jaredites to a new home in America, and the Book of Mormon records that they carried with them "Deseret," which Joseph Smith translated as "honey bee." The word "Deseret" was further mysterious because it is the only word in the Book of Mormon which survives from the ancient and holy language of Adam. This must have appeared to the immigrants to be an important key to the founding of the new Zion. The land of milk and honey, and the word "Deseret" and its visual symbol, the honey bee, gave great associational importance to the beehive emblems as the Mormons themselves fled from the cities of the unrighteous into the new land.

Deseret, represented by the working bee pioneers in the beehive Kingdom of God, became the most pervasive symbol in the building of the Great Basin Empire. This symbol, unlike the more formal and sacred icons of Grand Masonry, became a personal and exclusive symbol for the pioneer.

As the pioneers attempted to reorganize the hostile land according to an ideal inherited from the lush gardens of northern Europe, the beehive took on extra meaning. The hive as an agrarian model was crucial in the quest to reap the fruits of the earth and to make the desert "blossom as the rose." It is tempting to imagine that, given all the emphasis on the traditional skep beehive as a symbol, early Utah beekeepers wove skeps, invoked bee charms, and idolized their bees. The truth is that beekeepers in Utah used skeps only very early; they always employed the most scientific methods of beekeeping in order to produce as much as possible. The land must overflow with milk and honey.

The most important conception which the beehive suggested in those early days of settlement was the Christian belief that the beehive could be compared to "the Kingdom of God on Earth." This was interpreted as a communal life (the Mormons called their communal living "the United Order") with workers living a prescribed life of labor and with as few drones as possible. Brigham Young, as the king bee and the "all-seeing eye" of God, was the benevolent beekeeper.

As a political kingdom, the beehive was self-contained and was thus representative of the Mormon settlement. The beehive had definite boundaries and was completely protected. There were no windows looking out to the world and none looking into the Mormon kingdom. Mormons worried that the rest of the nation looked on them as attempting to replace the great tradition of American democracy with a perverted theocracy. And everything about the new land and its natives seemed hostile. The Mormons, suffering from what amounted to paranoia, treasured the idea that they were safe from both the outside world and their own new environment, and the beehive was a metaphor for this self-sufficiency.

The Council of Fifty, the earliest governmental body of Utah, unsuccessfully applied for statehood six times from 1849 to 1895. Each time Salt Lake City sent a constitution to Washington for the proposed State of Deseret, it gave up a little more autonomy, Mormon exclusiveness, and territory. Three years after the Mormons had entered the valley of the Great Salt Lake, the American eagle had swooped down and lighted protectively upon the beehive. In fact, the earliest material examples of the hive show it in conjunction with that other powerful political symbol, the eagle, common to both American and Masonic imagery. The Utah, not Deseret, Territory was created by the United States Congress, and Brigham Young was named governor. There continued to be a dual government for eighteen years: the legislature acted both as elected representatives of the Territory of Utah and as church-appointed workers for the religious State of Deseret. The use of the beehive and the word "Deseret" persisted during these years of conciliation and compromise. Most Mormon businesses of the early years employed the emblems of the beehive or the all-seeing eye, the words "Deseret" or "Zion," or the phrase "Holiness to the Lord" upon all signs, letterheads, and advertisements.

As Johnston's Army entered Salt Lake Valley in 1858 and the dream of an isolated kingdom ended forever, the symbols and rituals of the church began to take on new meaning. Secrecy shrouded the temple ceremony. Benevolent symbols such as the all-seeing eye looked too exclusive to outsiders and were resented by them. Such emblems ran against the grain of the American self-image: independent and free, the Americans wanted no eye, not even the eye of God, watching them. As statehood came closer, that foreign-sounding word, "Deseret," was dropped from all governmental use. The beehive also seemed too exclusive a symbol, but unlike the more formal and holy Mormon icons, the beehive was a fairly common figure throughout European and American tradition.

Like most folk symbols, the beehive, and with it the word Deseret, was adaptable to new times and circumstances. The politicians of the late nineteenth century looked at the beehive and contemplated a common meaning which pertained not only to Mormons, but to all people who would live in the state. The aspect of the beehive which suited this vision best was "industry," a virtue praised by all building civilizations and one particularly apt in the industrial age. So the beehive was allowed to stay. It was incorporated into the state seal; significantly, it was diminished in size, dwarfed by the threatening outstretched eagle. Instead of the proud "Deseret" banner flying above, the word "Industry" appeared.

Utah, Hive of Industry

Twentieth-century Utah, though still predominantly Mormon, found itself with the eclectic population which had been drawn to the crossroads of the West. The new industrious state used its shared symbol in a variety of ways; the beehive is employed widely and has come to carry a number of meanings, religious and secular. In Salt Lake City one finds a lavish Victorian hotel with fine china, linen, and decorative moldings all displaying the beehive beneath the eagle's talon. Price, originally a tiny Mormon village, experienced an early mining boom which brought in masses of gentile (non-Mormon) miners. Very early the Saints of Price incorporated the beehive into the city seal, making a statement of Mormon power amidst the coal-blackened faces of the foreigners. [Brigham City, a community of strong-willed farming Scandinavians, held tightly to the Mormon communal plan (the United Order) long after other communities

ceased adherence; the Box Elder High School Bees of Brigham City are just one organization which currently employs the bee and hive as mascot. In other parts of the state Mormon communities have resigned themselves to the irony of the fact that the beehive, emblem of Deseret, has become the official symbol for a state eventually named after a native tribe, the Utes, who happened to live within its boundaries.

In the early days of statehood, an urbane and slightly stinging periodical called *The Beehive* came and went. Samuel Auerbach, a Jewish merchant on the gentile side of the Salt Lake shopping district, offered a line of "Beehive Brand" clothing in 1916. "Deseret," in contrast with the wide and varied use of the beehive, remains a predominantly Mormon word.

The beehive as a religious, governmental, and commercial symbol has a long and varied history in Utah and is widely familiar to its citizens. From the variety of uses which we find it has been given, it seems clear that most craftsmen in early Utah were familiar with the motif and employed it in the course of their work with some regularity. Designers, stone carvers, wood turners, and sign painters all were prepared to craft versions of the skep shape, and thus we find the beehive a part of tombstones and gates, house eaves and leaded glass. Later, modern commerce extended its use to neon signs and letterheads, to souvenir honey pots and soda pop labels. The pithy social icon of the nineteenth century has shown itself to be an easily adaptable symbol

on the modern Utah scene. Its power, however, is much altered.

Once the beehive was an evocative symbol of
a fervently-held faith, capable of eliciting
sacrifice from believers and anger from their
enemies. Now it is a neutral traditional motif
and is used easily and widely, but at the cost
of much of its emotional force. The older
mystical symbols of Mormonism have been
replaced by sego lilies, handcarts, seagulls,
and the Mormon temple in silhouette.

The very neutrality of the beehive has made
available to us yet another incarnation of the
motif. Contemporary folk art and craft com-
bine the traditional with the new to bring
about a personal and creative use of the beehive. The variety of artistic expression the
beehive now permits is remarkable: the son of
a Navajo medicine man used the beehive in
a sand painting; a Tongan Mormon family im-
ports native materials to Utah in order to
create tapa cloth which uses the beehive motif;
an instrument-maker creates a banjo embel-
lished with the beehive in mother-of-pearl;
painters and sculptors are inspired by the
patterns which are part of the beehive shape.
All are working not only with the beehive
design but also with the ideas and history it
symbolizes. The beehive endures as an adaptable and lively folk symbol in Utah.

Suggested Reading

Arrington, Leonard J., and Davis Bitton. *The Mormon Experience: A History of the Latter-day Saints*. New York: Alfred A. Knopf, 1979.

Crane, Eva. *A Book of Honey*. New York: Charles Scribner's Sons, 1980.

Crane, Eva, ed. *Honey: A Comprehensive Survey*. New York: Crane, Russak, 1975. Published in cooperation with the International Bee Research Association.

Fife, Austin E. "The Concept of the Sacredness of Bees, Honey, and Wax in Christian Popular Tradition." PhD dissertation, Stanford University, Stanford, California, 1939.

Hansen, Klaus J. *Quest for Empire: The Political Kingdom of God and the Council of Fifty in Mormon History*. East Lansing, Mich.: Michigan State University Press, 1967.

Lambert, Asael Carlyle. "The Papers of Asael Carlyle Lambert" (MS 35). Manuscripts, Special Collections, University of Utah Library, Salt Lake City, Utah.

Lévi-Strauss, Claude. *From Honey to Ashes*. Translated from the French *Du Miel aux Cendres* by John and Doreen Weightman. New York: Harper and Row, 1973.

Morgan, Dale L. "Salt Lake City: City of the Saints." In *Rocky Mountain Cities*, ed. by Ray B. West, Jr. New York: W. W. Norton, 1949. Pp. 179-207.

Oman, Susan, and Richard Oman. "Mormon Iconography." In *Utah Folk Art: A Catalog of Material Culture*, ed. by Hal Cannon. Provo, Utah: Brigham Young University Press, 1980.

Ransome, Hilda M. *The Sacred Bee in Ancient Times and Folklore*. London: G. Allen and Unwin, 1937.

Roberts, Allen D. "Where Are the All-seeing Eyes: The Origin, Use, and Decline of Early Mormon Symbolism." In *Sunstone* (Salt Lake City, Utah), vol. 4, no. 3, pp. 22-37.

"But the Mormon crest was easy. And it was simple, unostentatious, and fitted like a glove. It was a representation of a Golden Beehive, with the bees all at work!"

Mark Twain, *Roughing It*

1, 2 Gravestones from (1) Wellsville, (2) Farmington

3 Gravestone from Ephraim

4   *Ancestral Image,* Meredith Gregg

His labor is a chant, His idleness a tune,
Oh for the bee's experience Of clover and of noon.   Emily Dickinson

5 Molded ice cream from Snelgrove's, which has offered the beehive mold for nearly a half-century. If there is a Mormon vice, it is ice cream.

''A bee-hive is the home of a swarm of bees. The swarm of bees receives its highest grade of perfection under the care of man—a bee-keeper.''

''The Beehive Symbolism'' in
L.D.S. Beehive Girls' Handbook
1925

6   Pioneer Day parade float, July 24, 1897, Salt Lake City. Photograph by Charles R. Savage

7   University of Utah seal in tile mosaic, Salt Lake City, c. 1912

8   Utah State Fairgrounds, Salt Lake City

9   Shoshone beaded purse, Wallace and Hazel Zundel, Clearfield, Utah

"Honey falls from the air
principally about the
rising of the stars, and
when the rainbow rests
upon the earth."

Aristotle

10   Pabst Beer mug, 1897

I see the gentleman from Utah, our friendly Beehive State.
How can we help you, Utah, how can we make you great?
Well, we got to irrigate in our desert, and we got to get some things to grow.
And we got to tell this country about Utah, because nobody seems to know.

Randy Newman, *"The Beehive State,"* 1969
© 1968 Six Continents Music Publishing, Inc., used by permission.

11   Totem pole, carved by Lyman S. Willardson, Logan, Utah, 1978

12 Newel-posts on the steps of the Mormon temple, St. George, Utah, c. 1877

"I will do my best to break down everything that divides. I will not have disunion and contention, and I mean that there shall not be a fiddle in the church, but what has 'Holiness to the Lord' upon it, not a flute, nor a trumpet, nor any other instrument of music."

Brigham Young,
c. 1845

13   Doorknob and escutcheon in the Mormon temple, Salt Lake City, c. 1890

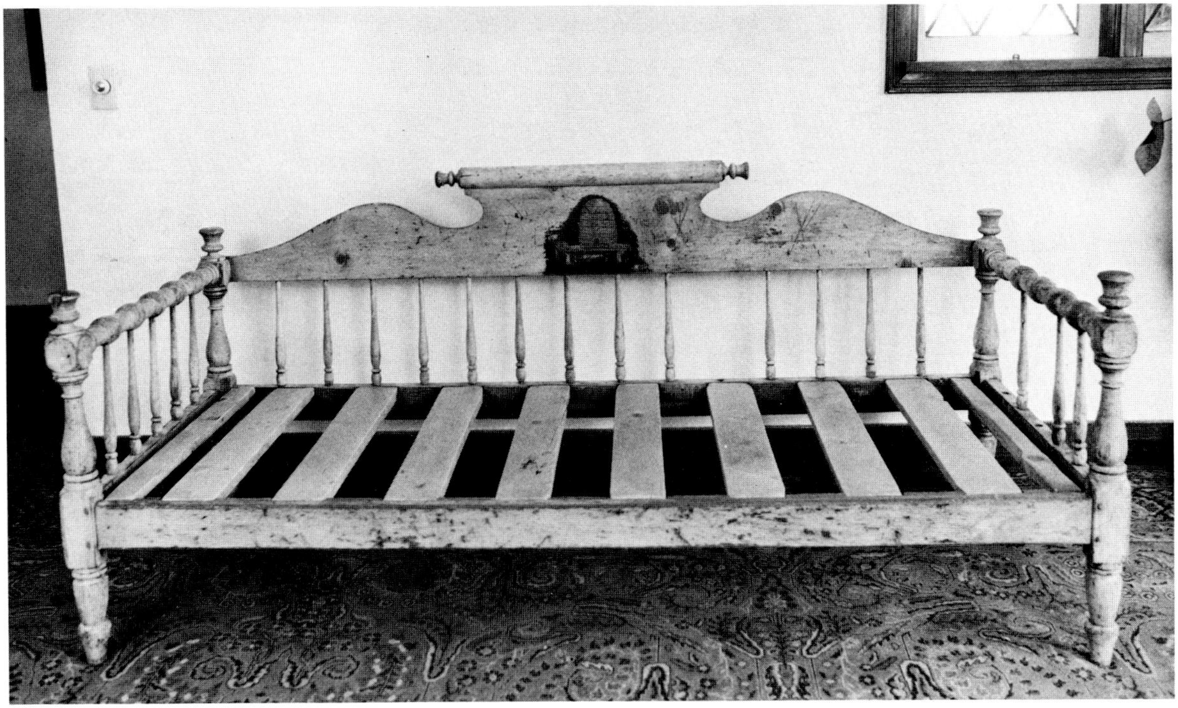

14   Utah pine lounge, made in northern Utah c. 1870's

15 Bedstead, carved by Ralph Ramsey. Ramsey, carver of the original Eagle Gate, was the most famous of the early Utah woodcarvers. The bedstead was begun in Salt Lake City in 1860 and completed in Mexico in 1898.

16   Patches and badges displaying the beehive

17   Manufacturers' and Merchants' Association label, Salt Lake City, 1906

''The fool seeketh to smite the bee, but the wise man leteth out the job to the lowest bidder.''

The Bee, 1898

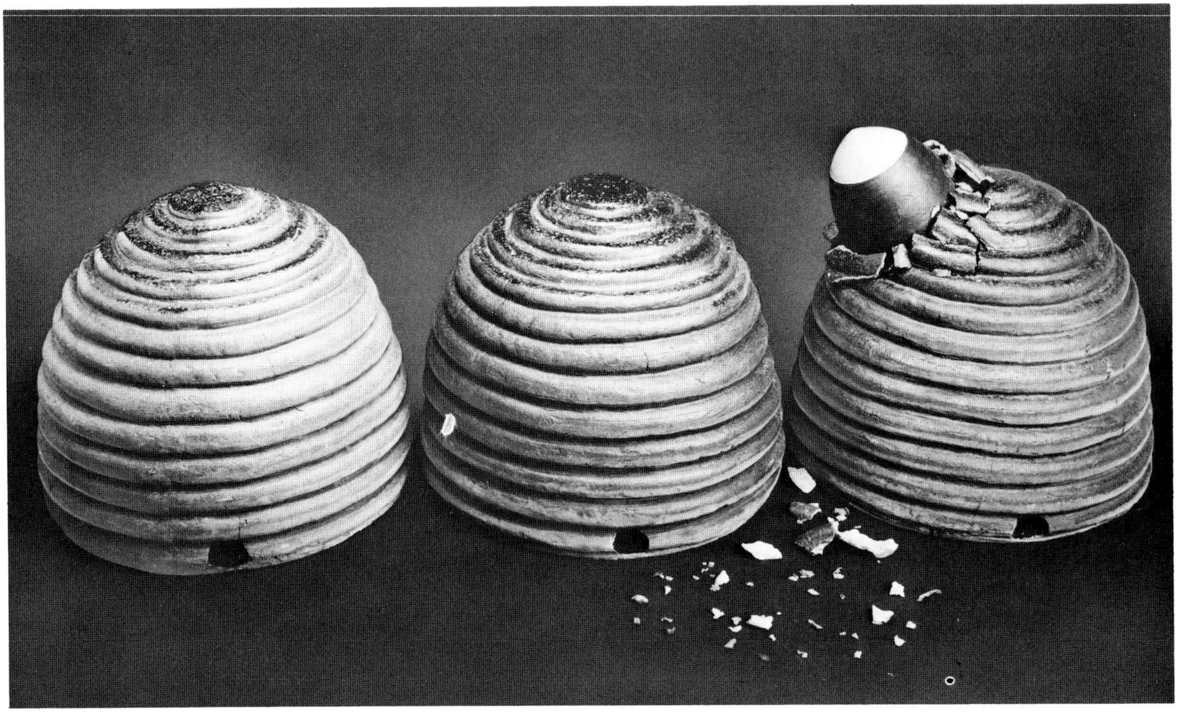

18 *Utah Shell Game*, David Pendell

"Uncle Sam. May he have strength in his old age to correct his unruly boys that they may cease pillaging the bee-hive or destroying the bees; that in times of winter they may have honey."

Toast offered on Pioneer Day,
July 24, 1851

19   Car doors with decals: Utah Highway Patrol, Price City Motor Pool, Gold Cross Ambulance, Utah State Motor Pool

20 Beehive Beverage Co., Ogden, Utah, one of over 150 businesses and organizations in Utah employing the beehive name

The busy bees of Deseret
    Are still around their hive,
Though honey-hunters in the world
    Don't wish these bees to thrive.

Ieuan, "The Bees of Deseret,"
published in *The Beehive Songster*, 1868.

21   Beehive Bakery, Salt Lake City. Fritz Haertel, ''the rye king,'' came to Utah early in this century from Chemnitz, Germany.

22   Eagle Gate and Beehive House, Salt Lake City, 1874

23 Eagle Gate and Beehive House, 1980

24   Attic vent of a turn-of-the-century house, Salt Lake City

"Deseret. The great Bee-hive. When the king of the Bees leads the way, all the Bees follow, and all are sure to obtain the honey, except the drones and they are left by the way."

Toast offered on
Pioneer Day,
July 24, 1851

25   Fire alarm pedestal, Salt Lake City

"When God was distrib-
uting gifts to the animals
the bees asked to be
lodged in golden baskets
and that their sting be
always deadly, so as to be
respected therein. God,
to chastise them for their
arrogance, lodged them
in straw skeps and caused
that a bee should die
whenever it stings."

A Christian fable

26   *Land of Milk and Honey*, John Shaw and Diane Shaw

27   Square dancers, Salt Lake City. The beehive costume was used for the 1973 National Square Dance Convention.

29 *Beehive*, Frank A. Smith

30 Deseret tapa cloth, Ilaese Lavulavu and Lita Mulitalo, Salt Lake City. The design combines the Tongan coat-of-arms with the Mormon beehive.

"The hive and honey bee form our communal coat of arms. The symbol is adopted extensively in our local institutions. It is a significant representation of the industry, harmony, order and frugality of the people, and of the sweet results of their toil, union and intelligent cooperation."

Deseret News, 1881

31   Centennial rug, Salt Lake City. In 1947 the Ouzounians, an Armenian family of Mormon converts, made this rug in their home in Syria for the centennial of the Mormons' arrival in the Salt Lake valley.

32 Daughters of Utah Pioneers Museum, Salt Lake City

33   Liberty Ward chapel, Salt Lake City

''Our meeting was well attended and manifested an increased desire to make Utah a success in the bee culture until this land shall flow with honey as it already does with milk.''

Edward Stevenson, Secretary
Utah Bee Keepers Conventions, 1882

34　Christian Otteson, beekeeper, Huntsville, Utah. Photograph by George E. Anderson, c. 1900

35 Skep beehive of woven straw, Holland, contemporary

36   Over one hundred artists contributed postcard versions of the beehive theme to a correspondence
art project created by David S.

Disregard the location of 50 beehives from memory
H. TRAILOR SELGERS
UNITED STA
TAX COLLECTOR   STATE OF UTAH

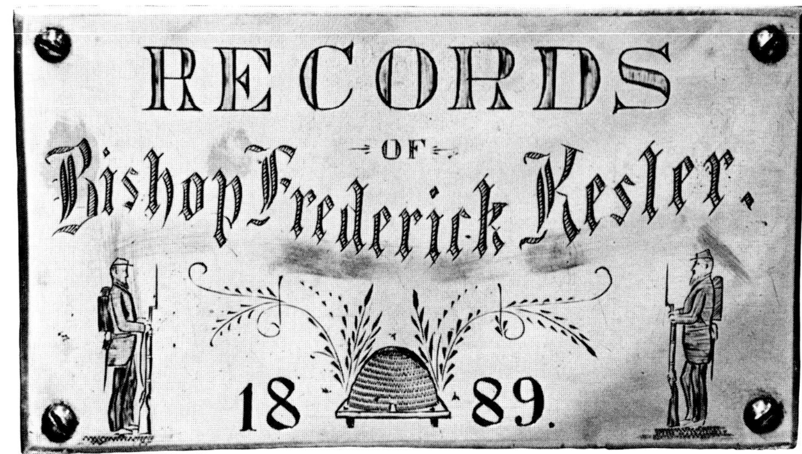

38 Engraved bronze plate, records chest, Salt Lake City, 1889

39 Zinc coffin plate, c. 1860

How doth the little busy bee
Improve each shining hour
And gather honey all the day
From each opening flower.

Isaac Watts, seventeenth-century hymn

40　Engraved sterling silver plate, records chest, Salt Lake City, c. 1889

Oh, Brigham said, "The Mormons
Should be busy as the bees
And build a home in Utah
Where they could all be free
To practice their religion
Apart from all the rest."
Oh, to hear the busy humming
Of the bees in Deseret, boys,
Oh, to hear the busy humming
Of the bees in Deseret.

Mark Jardine, *"Oh, Deseret"*

41   Deseret banjo headstock, Leonard Coulson, Intermountain Guitar and Banjo, Salt Lake City

42 Cake from Mrs. Backer's Pastry Shop, Salt Lake City. Martin Backer's flair for baking detailed thematic cakes has led to calls for many beehive cakes during his 40 years of baking.

43 Letterhead logotypes from Utah businesses employing the beehive

"The Beehive State. May her valley cells be filled with the honey of her own production, and the bees seek from the flowers of the Valley that which makes them independent in their rainy days…"

Toast offered on Pioneer Day, July 24, 1851

44 *Deseret Alphabet First Book*, cover. In 1868 leaders of the Mormon church experimented for a short time with a new phonetic alphabet.

45 Salt and pepper shakers

46 *Beehive*, Lee Deffebach

"That which is not good
for the swarm, neither
is it good for the bee."
Marcus Aurelius

47   *The Book of Life*, Alfred Raymond Wight, 1949. This pine carving
symbolically illustrates the Mormon plan of eternal progression.

48   *Beehive Quilt*, Harry Taylor

49　The Hotel Utah at Christmas in the 1950's, Salt Lake City

''The tower, rising four stories above the roof garden, is capped with a mammoth beehive set on a pedestal framing on each of its four panels the eagle and shield, and the words 'Hotel Utah' which, in colored lights at night can be read for many miles. The pedestal is twenty-two feet square, the beehive is twenty-two feet high and sixty feet in circumference at its base.''

Hotel Monthly, 1911

50 The Mormon temple seen from the Hotel Utah's Roof Restaurant

51 Banner of the Eighteenth Ward of the Mormon church, Salt Lake City

"I now will tell you something
  You never thought of yet;
We bees are nearly filling
  The hive of Deseret:
If hurt, we'll sting together,
  And gather all we get;
    For all are talking of Utah."

Ieuan, "All Are Talking of Utah," published in *Beehive Songster*, 1868

52 Endowment House, Spring City, Utah, 1876

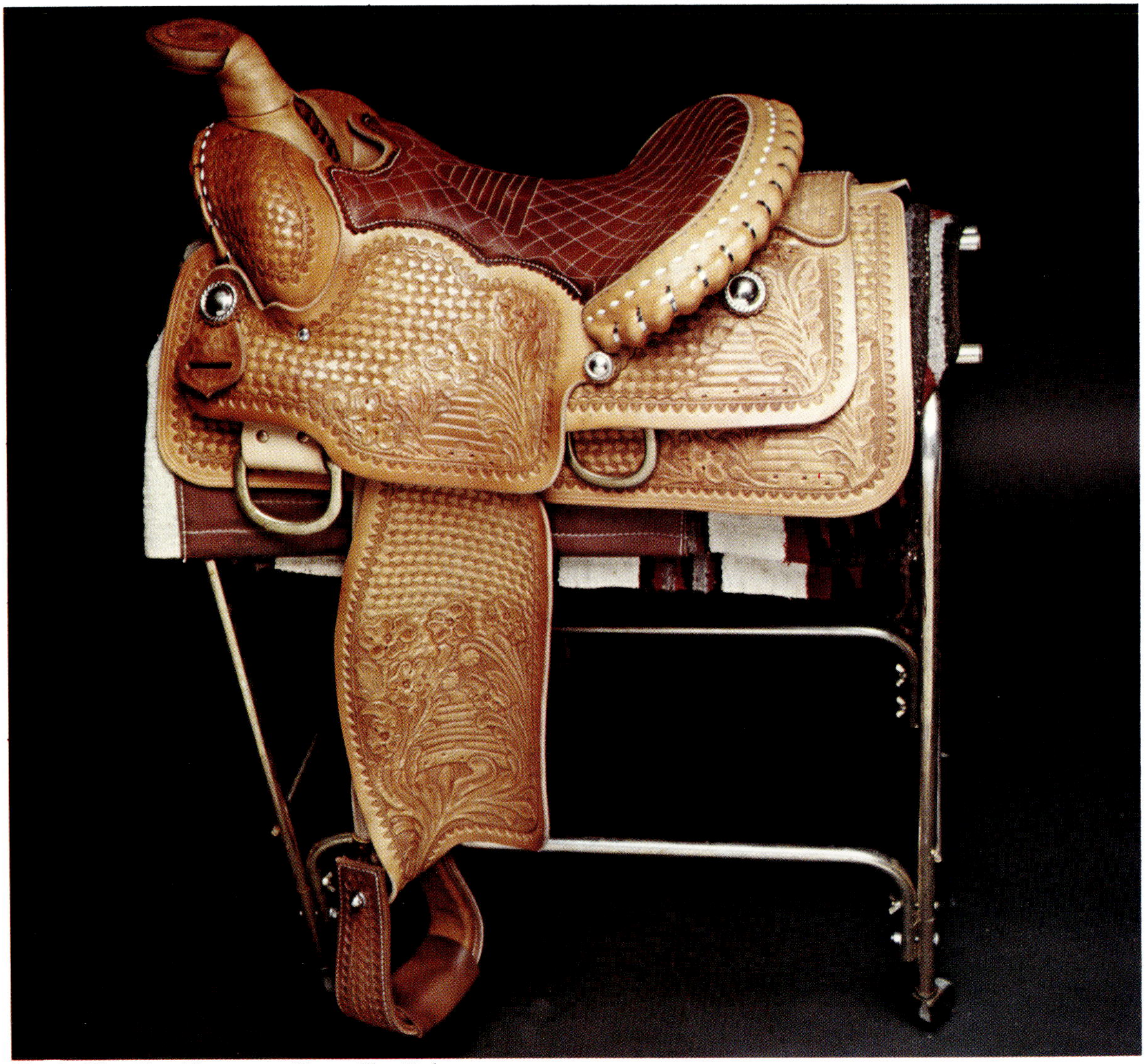

53   Saddle, Glen Thompson, Huntsville, Utah

"When Ra weeps again the water. which flows from his eyes upon the ground turns into working bees. They work in flowers and trees of every kind and wax and honey come into being."

Early Egyptian reference from the Salt Magical Papyrus

54 *Landmark, Wasp War Memorial,* and *Tourist Trap,* Maureen O'Hara Ure

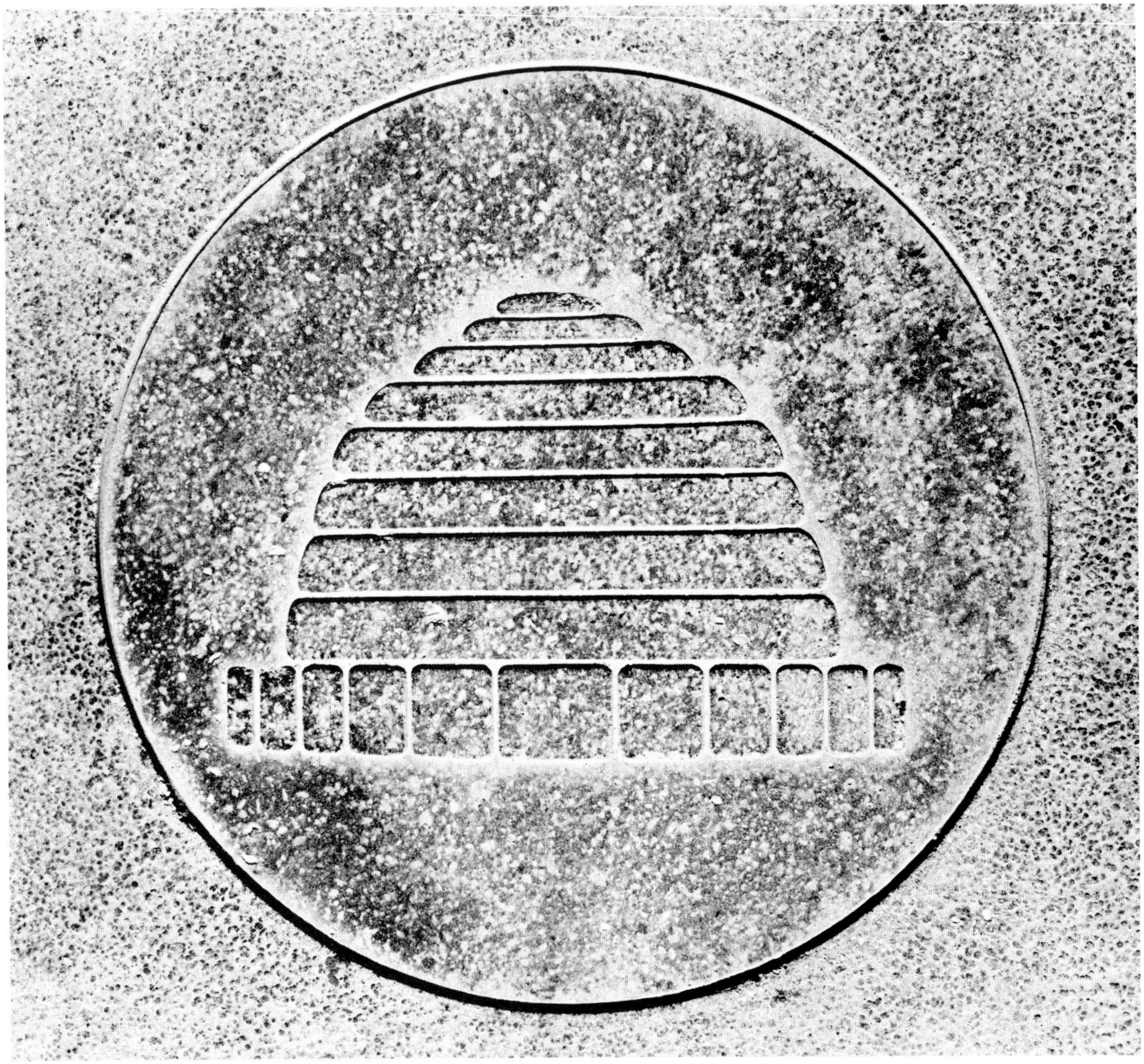

55   Sidewalk, Salt Lake City

56　*The Beekeepers*, from the workshop of Pieter Brueghel the Elder, 1565

"He that medleth…with the bees, must specially keepe himselfe from letchery,
and drunkennesse, and wash himselfe cleane: for they love to have such as come
about them to be as pure and clean as may be."

—from a seventeenth-century
English agricultural manual

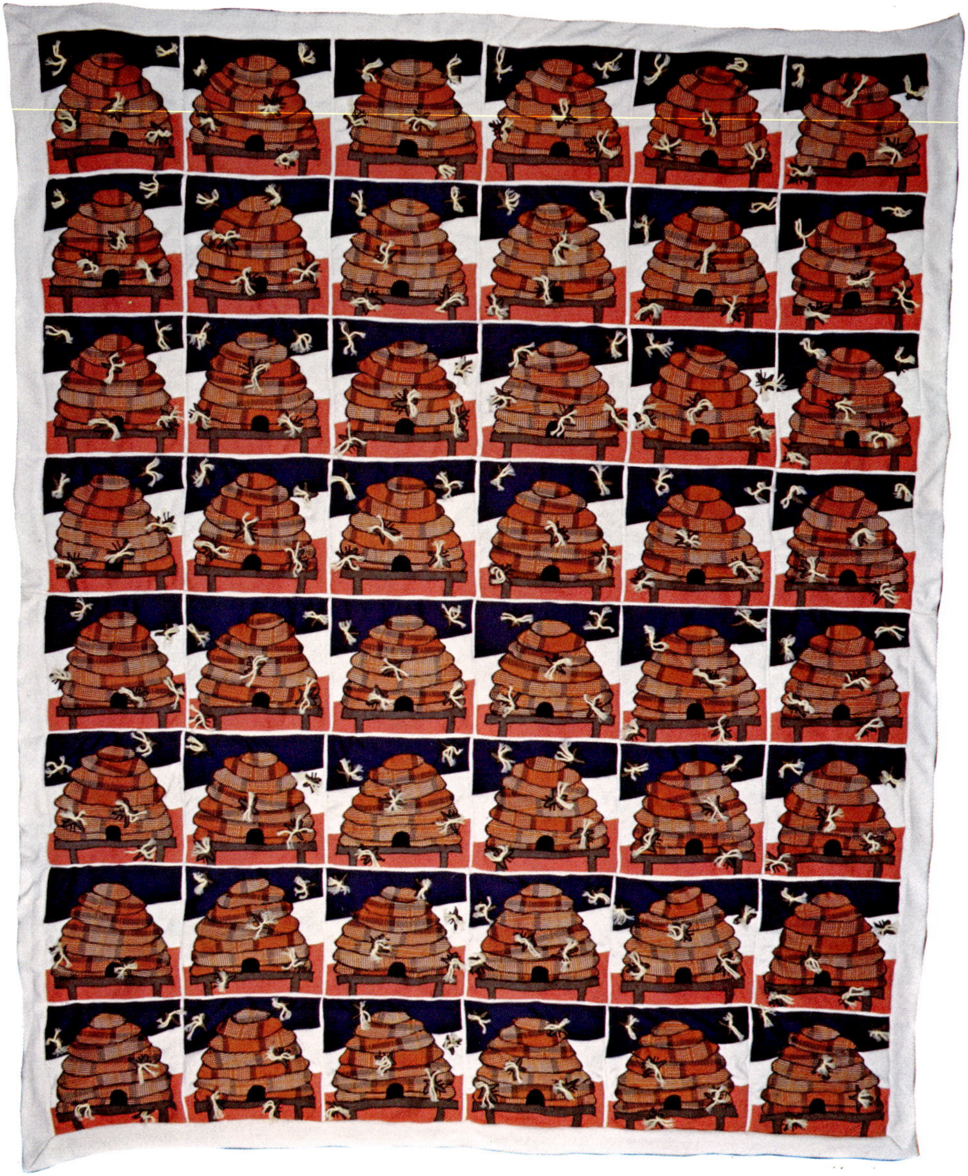

57   Quilt, Bud King, Salt Lake City

58 .Blanket, Utah Woolen Mills, Salt Lake City, 1927

59   Representation of Christ holding a beehive, Tur Büttgens, Austrian, 1927

"Conspicuously stands the Rush Bee-hive braided whole, with the bees swarming about sucking honey from the morning glory whose blooming vines entwine…. The all-seeing eye seems watching the busy scene…."

Brigham Young, 1852

60   Utah's Washington Monument block, carved by William Ward. Each state and territory contributed a block for the construction of the Monument in 1852.

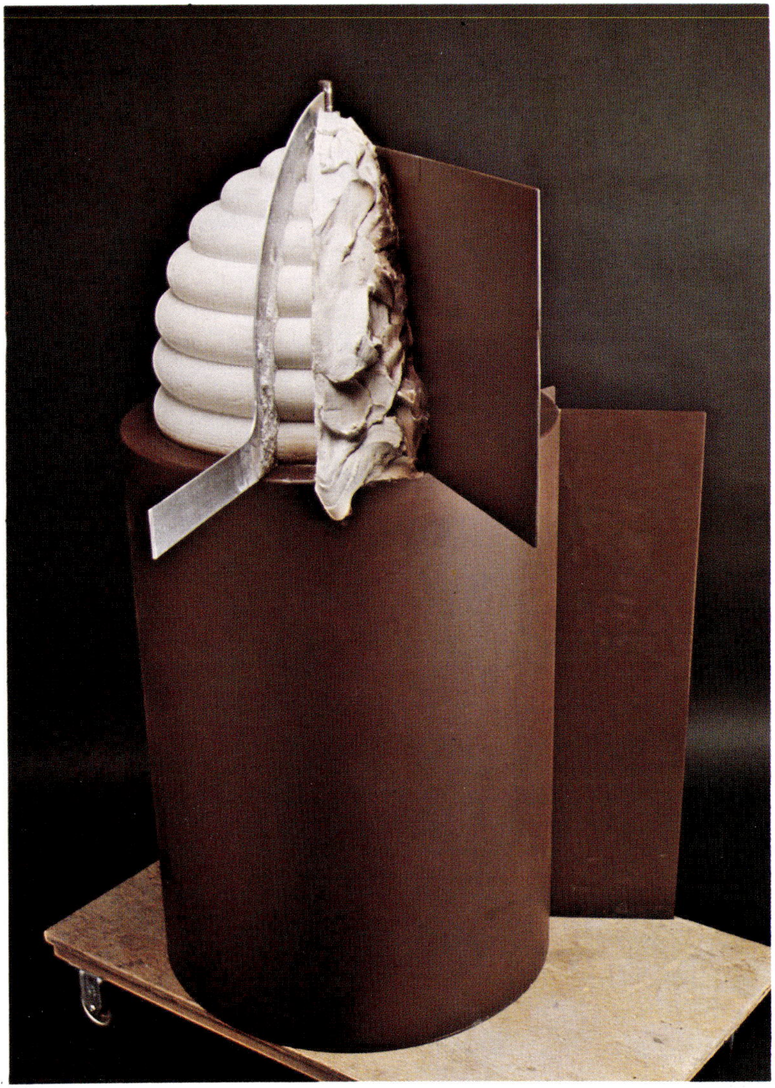

61 *Significantly White and Incomplete*, Richard Johnston

62 *Self-portrait with Beehive: Reflections upon Becoming a Resident of Utah*, Lou Ann England

63   Sand painting by Joe and Victoria Kaibetoney, Salt Lake City. The sand painting has a sacred tradition. To a new generation of Navajos, it has become as well a popular decorative art.

64   Steeple clock, made in Connecticut c. 1880

65 Beehive Cleaning Co., Salt Lake City

66   *Utah, the Beehive State*, Bonnie Sucec

67 Box Elder High School Color Guard

68 Utah state highway signs, Farmington, Utah

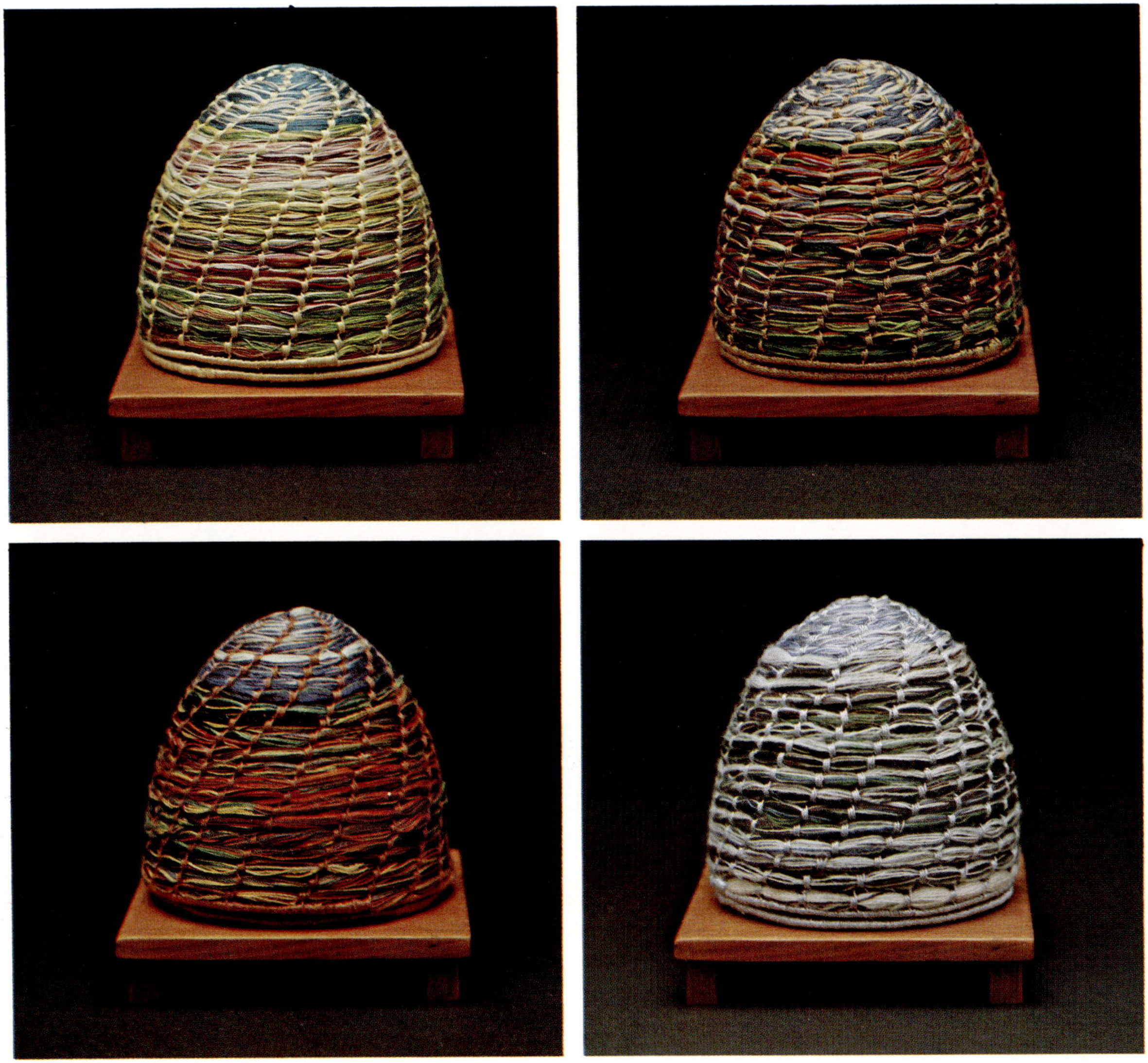

69   *Four Seasons of Deseret*, Ann C. Pendell

70   Wrought iron fence, Provo, Utah

Plates

1. GRAVESTONE (Ira Ames); 1869; Wellsville Cemetery, Wellsville, Utah; pink sandstone, 31″ high, 17″ wide; fiber glass facsimile lent by the Utah Arts Council, executed by Paul Reynolds, 1978.

2. GRAVESTONE (David R. Roberts); 1858; Farmington Cemetery, Farmington, Utah; tan sandstone, copper plate; 36″ high, 20″ wide.

3. GRAVESTONE (Anders Jensen); 1872; Old Ephraim Cemetery, Ephraim, Utah; tan sandstone; 41½″ high, 23½″ wide.

4. "ANCESTRAL IMAGE"; 1980; Meredith M. Gregg, Salt Lake City, Utah; fabric, paper, ink, prismacolor, xerox, pastel, acrylic; 22″ high, 28″ wide; lent by Meredith M. Gregg.

5. MOLDED ICE CREAM; c. 1930—; Epelscheimer & Co., New York; lead alloy mold, ice cream; 2¾″ high, 7″ circum.; courtesy of Snelgrove Ice Cream Co.

6. "PIONEER DAY FLOAT"; 1897; Charles R. Savage, Salt Lake City, Utah; photograph courtesy of Church of Jesus Christ of Latter-day Saints Historical Department.

7. TILE MOSAIC; c. 1912; John R. Park Building, University of Utah, Salt Lake City, Utah; 72″ high, 72″ wide; photograph courtesy of the University of Utah.

8. STATE FAIRGROUNDS ENTRANCE; 1966; Salt Lake City, Utah; fiber glass, wire, wood, plexiglass, neon; 96″ high, 120″ wide; courtesy of the Utah State Division of Expositions.

9. SHOSHONE PURSE; 1980; Wallace Zundel and Hazel Zundel, Layton, Utah; buckskin, glass beads; 21½″ high, 9½″ wide; lent by the Utah Arts Council.

10. MUG; 1897; Pabst Brewing Co., Milwaukee, Wisc.; ceramic; 4¼″ high, 11″ circum.; lent by M.A. Whitelock.

11. TOTEM POLE; 1978; Lyman S. Willardson, Logan, Utah; pine; 85½″ high, 22″ circum.; lent by Alice Willardson.

12. NEWEL-POSTS; c. 1877; Truman O. Angell, architect; Church of Jesus Christ of Latter-day Saints Temple, St. George, Utah; sandstone, whitewash.

13. DOORKNOB & ESCUTCHEON; c. 1890; designed by Joseph Don Carlos Young, Church of Jesus Christ of Latter-day Saints Temple, Salt Lake City, Utah; brass; 18″ high, 4″ wide; photograph lent by the Corporation of the President of the Church of Jesus Christ of Latter-day Saints.

14. UTAH LOUNGE; c. 1870's; northern Utah; pine; 32″ high, 74½″ wide; lent by Stephanie Wilde.

15. BEDSTEAD; 1860-1898; Ralph Ramsey, Utah, Arizona, Mexico; box elder, cedar, red cedar, pine; footboard 52″ wide, 32″ high; courtesy of the Daughters of Utah Pioneers Museum, Salt Lake City, Utah.

16. PATCHES & BADGES; c. 1900-1980; Salt Lake City, Utah; 4″ high, 3½″ wide to 1″ high, 1″ wide; private collection.

17. LABEL (Manufacturers' & Merchants' Association); 1906; Utah; 1½″ high, 2½″ wide; courtesy of Utah State Archives.

18. "UTAH SHELL GAME"; 1980; David Pendell, Salt Lake City, Utah; stoneware; each piece 26″ high, 84″ circum.; lent by David Pendell.

19. CAR DOOR DECALS; contemporary; Utah; vinyl plastic; 11½″ high, 16″ wide to 10½″ high, 8″ wide; lent by Gold Cross Ambulance Service, Price City Motor Pool, Utah Highway Patrol; Utah State Motor Pool.

20. SODA BOTTLES AND CASE; c. 1945-1970; Beehive Bottling Co., Brigham City, Utah; 10″ high, 18½″ wide; lent by Myron Barlow, Seven-Up–Dr. Pepper Bottling Co.

21. NEON SIGN (Beehive Bakery); c. 1950; Gordon Smith, Rainbow Sign Company, Salt Lake City, Utah; 72″ high, 72″ wide; lent by Fritz Haertel.

22. "BEEHIVE HOUSE & EAGLE GATE"; c. 1874; Salt Lake City, Utah; photograph courtesy of the Utah State Historical Society.

23. "BEEHIVE HOUSE & EAGLE GATE"; 1980; Brent Herridge, Salt Lake City, Utah; photograph.

24. ATTIC VENT; c. 1892; Salt Lake City, Utah.

25. FIRE ALARM PEDESTAL; c. 1900; Gamewell Co.; steel, paint; 60″ high, 45″ circum.

26. "LAND OF MILK & HONEY"; 1980; John Shaw & Diane Shaw, North Salt Lake, Utah; clay, metal, oak; 36″ high, 24″ wide; lent by John Shaw & Diane Shaw.

27. SQUARE DANCING COSTUMES; 1973; Leone Conrady, Salt Lake City, Utah; cotton; lent by Leone Conrady, modeled by Bob Young & Penny Young.

28. HONEY POTS; 1900-1980; Utah; ceramic; 5″ high, 17½″ circum. to 3½″ high, 13½″ circum.; private collections.

29. "BEEHIVE"; 1980; Frank Anthony Smith, Salt Lake City, Utah; acrylic on paper; 27″ high, 23½″ wide; lent by Frank Anthony Smith.

30. TAPA CLOTH; 1980; Ilaese Lavulavu & Lita Mulitalo, Salt Lake City, Utah; mulberry fibers, raffia, shells, pods; 48″ high, 57½″ wide; lent by the Utah Arts Council.

31. RUG; c. 1947; Ruben & Mary Ouzounian, Syria; cotton, silk, wool, dye; 25½″ high, 18″ wide; courtesy of the Daughters of Utah Pioneers Museum, Salt Lake City, Utah.

32. STAINED GLASS WINDOW; c. 1950; adapted from design by A. Fairbanks, Utah, executed by J.N. Lucas, Church Art Glass Studio, San Francisco; leaded, colored, & painted glass; 110″ high, 80″ wide; courtesy of the Daughters of Utah Pioneers Museum, Salt Lake City, Utah.

33. STAINED GLASS WINDOW; c. 1908; Europe; painted and leaded glass; 108″ high, 78″ wide; courtesy of Liberty Ward.

34. "CHRISTIAN OTTESON, BEEKEEPER"; c. 1900; G.E. Anderson, Huntsville, Utah; photograph courtesy of George Edward Anderson Collection, Brigham Young University Library.

35. SKEP; contemporary; Holland; woven straw; 14½″ high, 48″ circum.; lent by Miller Honey Co.

36, 37. CORRESPONDENCE ART PROJECT; 1980; international; director: David S. (Sucec), Salt Lake City, Utah; mixed media; each card 3½″ high, 5½″ wide; contributed by S. Cooper, H. Gregor, R. Lake, C. Miller, A. Mullin, Musicmaster, C. Nagasawa,

M. Price, H.F. Sellers, F.A. Smith, D.J. Solomon, M.O. Ure.

38. RECORDS CHEST PLATE (interior); 1889; designed by Frederick Kesler, Salt Lake City, Utah; engraved bronze; 3½″ high, 6″ wide; courtesy of Special Collections, University of Utah Library.

39. COFFIN PLATE; c. 1870's; Sargent & Co., New Haven, Conn.; zinc alloy; 3″ high, 4″ wide; private collection.

40. RECORDS CHEST PLATE (exterior); 1889; designed by Frederick Kesler, Salt Lake City, Utah, cast in Philadelphia, Pa.; engraved sterling silver; 5″ high, 6″ wide; courtesy of Special Collections, University of Utah Library.

41. BANJO; 1980; Leonard Coulson, Salt Lake City, Utah; maple, ebony, mother-of-pearl, gold plate, brass, abalone; 40″ high, 14½″ wide; lent by Leonard Coulson.

42. CAKE; 1980; Martin A. Backer of Mrs. Backer's Pastry Shop, Salt Lake City, Utah; butter cream & royal icing; 11″ high, 19″ wide at base; lent by the Utah Arts Council.

43. LETTERHEAD LOGOTYPES; 1980; Utah; contributed by Beehive Appliance Service, Beehive Cleaning Co., Beehive Clothing Mills, Beehive Golf Club, Beehive International, Beehive Karting & Recreation, Beehive Machinery, Brigham Young University, Deseret Federal Savings & Loan.

44. COVER, "DESERET ALPHABET FIRST BOOK"; 1868; New York; 7½″ high, 5″ wide, lent by the Utah Arts Council.

45. SALT & PEPPER SHAKERS; 1950-1980; Utah; ceramic; 5″ high, 9½″ circum. to 2½″ high, 7″ circum.; private collections.

46. "BEEHIVE"; 1980; Lee Deffebach, Salt Lake City, Utah; acrylic, oil, mixed media on canvas; 67″ high, 61″ wide; lent by Lee Deffebach.

47. "THE BOOK OF LIFE"; 1949; Alfred Raymond Wight, Draper, Utah; wood, light fixtures; 41″ high, 30″ wide; lent by the Church of Jesus Christ of Latter-day Saints Historical Department.

48. "BEEHIVE QUILT"; 1980; Harry G. Taylor, Ogden, Utah; photo etching, 17″ high, 14″ wide; lent by Harry G. Taylor.

49. "HOTEL UTAH AT CHRISTMAS"; 1950's; Hal Rumel, Salt Lake City, Utah; photograph courtesy of the Hotel Utah.

50. CHINA, GLASSWARE, FLATWARE; 1911-1980; 11" plate, Scammell Co.; amber goblet, Libby Co., silverware, Reed & Barton; liqueur glass; lent by the Hotel Utah.

51. POLITICAL BANNER; date unknown; Eighteenth Ward, Salt Lake City, Utah; velvet, silk, gold braid; 54" high, 39½" wide; courtesy of the Daughters of Utah Pioneers Museum, Salt Lake City, Utah.

52. SYMBOLIC INSCRIPTION STONE; 1876; Endowment House, Spring City, Utah; oolitic limestone, 18" high, 60" wide; courtesy of Allen D. Roberts.

53. SADDLE; 1980; Glen Thompson, Huntsville, Utah; leather, rawhide, trim; 40" high, 28" wide; lent by Glen Thompson.

54. "THE LANDMARK" (5" high, 7" wide), "WASP WAR MEMORIAL" (4½" high, 5" wide), "TOURIST TRAP" (5½" high, 9" wide); 1980; Maureen O'Hara Ure, Salt Lake City, Utah; found objects, acrylic, ink, pencil; lent by Maureen O'Hara Ure.

55. SIDEWALK; 1970; designed by Kevin Watts, Snedaker, Budd, & Watts, Architects, Salt Lake City, Utah; concrete; 34½" high, 34½" wide.

56. "THE BEEKEEPERS"; c. 1565; from the workshop of Pieter Brueghel the Elder, Northern Europe; photograph of original pen and ink rendition; original drawing 8" high, 12" wide; photograph courtesy of the Achenbach Foundation for Graphic Arts.

57. QUILT; 1980; Bud King, Salt Lake City, Utah; polyester fibers; 92" high, 74" wide; lent by the Utah Arts Council.

58. BLANKET; 1927; Utah Woolen Mills, Salt Lake City, Utah; double weave wool; 72" high, 54" wide; lent by the Church of Jesus Christ of Latter-day Saints Historical Department.

59. REPRESENTATION OF CHRIST HOLDING A BEEHIVE; 1927; Tur Büttgens, Austria; 20" high, 15" wide; courtesy of the Office of Apiculture, Cornell University.

60. WASHINGTON MONUMENT BLOCK; 1852; carved by William Ward, Salt Lake City, Utah; oolitic limestone, 24" high, 36" wide; daguerreotype courtesy of Church of Jesus Christ of Latter-day Saints Historical Department.

61. "SIGNIFICANTLY WHITE & INCOMPLETE"; 1980; Richard Johnston, North Salt Lake, Utah; steel, plaster; 60" high, 76" circum.; lent by Richard Johnston.

62. "SELF-PORTRAIT WITH BEEHIVE: REFLECTIONS UPON BECOMING A RESIDENT OF UTAH"; 1980; Lou Ann Heller England, Salt Lake City, Utah; oil on canvas; 39" high, 48" wide; lent by Lou Ann Heller England.

63. NAVAJO SAND PAINTING; 1980; Joe Kaibetoney & Victoria Kaibetoney, Salt Lake City, Utah; aspen, sand; 9" high, 23" wide; lent by the Utah Arts Council.

64. STEEPLE CLOCK; 1880's; Connecticut; rosewood veneer, reverse painting; 18" high, 10½" wide; lent by Stephanie Wilde.

65. NEON SIGN (Beehive Cleaning Co.); c. 1945; designed by Floyd Bueter, Salt Lake City, Utah; 42" high, 84" wide; courtesy of Gerald Bueter.

66. "UTAH, THE BEEHIVE STATE"; 1980; Bonnie Sucec, Salt Lake City, Utah; wood, dough, gouache, acrylic, wire; 36" high, 48" wide; lent by Bonnie Sucec.

67. COLOR GUARD UNIFORMS; 1980; Box Elder High School, Brigham City, Utah; wool, felt; lent by Jan Hondzinski, modeled by Lisa Kellogg and Mindy Webster.

68. HIGHWAY SIGNS; c. 1960—; Farmington, Utah; 24" high, 22" wide; lent by the Utah State Department of Transportation.

69. "FOUR SEASONS OF DESERET"; 1980; Ann Chalmers Pendell, Salt Lake City, Utah; wood, rawhide, embroidery floss; each piece 5" high, 12½" circum.; lent by Ann Chalmers Pendell.

70. WROUGHT IRON FENCE; 1890's; Provo, Utah; each section 37" high, 57" wide; lent by the Norma Giles Thomas family.

FRONT COVER: detail of "HOTEL UTAH AT CHRISTMAS" (see plate 49).
BACK COVER: detail of STAINED GLASS WINDOW (see plate 33).